A Program on Construction Ethics

A Publication of the American Institute of Constructors

I0392628

CAUTION REGARDING ANTI-COMPETITIVE BEHAVIOR
Compliance with antitrust laws is an indispensable element of ethical conduct in the construction industry. Construction contractors and others should be mindful of the possibility that discussion of the case studies on ethical relationships contained herein could, conceivably, lead to discussions of agreements that illegally restrain competition.

Sponsors, moderators of, and participants in construction ethics programs should ensure that discussions on the case studies do not in any way diverge into forbidden topics, such as would promote anti-competitive bidding procedures, pricing or bidding structures or agreements not to deal with particular firms or types of firms. To do otherwise could create a risk of antitrust liability. This is of particular importance with regard to case studies 2, 4, 7, 11, 15-17, 19,21,23, 28, 28-30, 30, and 33, which relate to competitive practice issues.

ABOUT THE AMERICAN INSTITUTE OF CONSTRUCTORS

Founded in 1971, the American Institute of Constructors mission is to promote individual professionalism and excellence throughout the related fields of construction. AIC supports the individual Constructor throughout their careers by helping to develop the skills, knowledge, professionalism and ethics that further the standing of the construction industry. AIC Members participate in developing, and commit to, the highest standards of practice in managing the projects and relationships that contribute to the successful competition of the construction process. In addition to membership, the AIC certifies individuals through the Constructor Certification Commission. The Associate Constructor (AC) and Certified Professional Constructor (CPC) are internationally recognized certifications in the construction industry. These two certifications give formal recognition of the education and experience that defines a Professional Constructor. For more information about the AIC, visit www.professionalconstructor.org.

AIC MEMBERSHIP

Become a member of the American Institute of Constructors by visiting our website at www.professionalconstructor.org. Membership within the AIC provides individuals with the tools and resources needed to grow as a construction professional. AIC Members enjoy access to unique benefits and valuable discount programs as well as a wide range of professional education opportunities. Membership is open to individuals ranging from Students to Professionals.

Contents

FORWARD

Few professional endeavors involve so many participants, handling so much information, under such severe competitive pressure as management of the construction process. Every constructor must have a well-developed sense of ethical conduct in order to perform successfully in this environment.

The American Institute of Constructors believes that its STANDARDS OF PROFESSIONAL CONDUCT are more than simple words. Indeed, the practice of true professionalism entails a commitment to ethical conduct, and ethics education is an important part of continuing professional development.

The following materials are offered in support of traditional AIC programs developed by and for professional constructors. It is our common goal to foster growth in the understanding, acceptance, and application of professional ethical conduct throughout the construction industry.

OBJECTIVE

This program is directed, primarily, toward constructors and AIC student chapter members. However, it should be of interest to and have applications for all construction program students and all individuals who regularly deal with contractors or who are otherwise directly involved in the construction industry. The purpose of this program is:

- To create an understanding of construction ethics.
- To provide an awareness of the ethical questions that arises in normal construction business operations.
- To develop an understanding of the importance (consequences) of one's actions (on our firms, our industry, ourselves, and others) when making ethical decisions.
- To create a better industry and environment in which to work.

This is not a program for those seeking answers to specific ethical questions. This program does not, nor is it intended to, provide clear "Yes" and "No" answers. Such decisions are for the participants to derive – if they can.

INTRODUCTION

This PROGRAM ON CONSTRUCTION ETHICS has been developed to provide AIC chapters, AIC student chapters, constructors and participants with simple guidelines to conduct a 60-minute to two-hour program on the subject of applied personal/professional and business ethics for the issues arising in the construction industry. The purpose is also to present the materials in this manual to the university community who are teaching students in various construction programs including Construction Management, Construction Engineering and Construction Technology. The concepts for discussion are within the personal and general societal thinking and conduct assumed and expected within historical western moral values.

The goals of the program are:

- To create an awareness and teach an understanding of ethics for the construction industry.
- To offer case studies as ethical questions arising in the normal course of construction business operations.
- To develop an understanding of the importance of one's decision-making process and resulting actions (consequences) on all the participants of our industry including our employees and ourselves as constructors.
- To create, enhance and maintain a better industry and an environment within which to work that enhances the common good for all.
- To join with full knowledge and commitment to the implementation of the AIC Constructor Code of Conduct.

It is not the goal of this material to have students, constructors and participants bogged down in philosophical terms involving the history or theory behind "ethics". Nor is it the goal of this material to require conformity to one person's view of morality as opposed to that of another person. However, it is the goal of this material to encourage students, constructors and participants within the construction industry to choose to be positively shaped into the ethical leaders of tomorrow for the construction profession. Further, it is the goal of this program to develop committed ethical leaders what will lead our construction industry to a more recognized prominence in achieving, supporting and maintaining the common good for all within our industry.

"Good ethics is good business!" We have all heard this saying, but is it really true? What is ethics? What is the purpose of ethics in our lives and/or our businesses? How do we get ethics? What do we do with the ethics that we learn and acquire? These questions strike at the heart of our knowledge and understanding of ethics and applied ethics and must be addressed if we are to understand and function in an ethical environment. The answers to these questions must enter into our decision-making process. Ethics, by its very definition, requires moral deliberation about questions raised from conflicting choices in our lives and businesses.

Let me state at the outset that there are no ethics in the construction profession. There are only ethical people in the construction profession. Professions cannot make moral decisions or deliberate over moral principles, only people can do so. Therefore, it is people who must address their ethics. People must choose their decision-making methods. Then people of common commitment can brand together for the common good of all persons and become leaders in applying their common chosen moral foundations with their profession.

WHAT IS ETHICS?
Most definitions of ethics recognize the moral element. Generally, it can be said that ethics is an individual's discovered and chosen moral code or self-standard, which is used within the decision-making process to resolve conflicts within his/her life or business. In the choices made daily in our lives and business we make evaluations that affect our decision-making process and the resulting decisions. These decisions affect people and the world around us as well as ourselves. Ethics is that moral element that deals with "what is right and what is wrong"

while we are in the decision-making process. Or, "what I can do versus what I should do." Ethics can be considered the moral filter through which other considerations such as people, economics, environment and law, etc. are placed within the larger view of good and evil or, perhaps, the greater good when there are competing goods.

WHAT IS THE PURPOSE OF ETHICS IN OUR LIVES AND OUR BUSINESSES?

There are a number of views to this question just as there are too many ethical questions. Purposes may be noble, that is, to assure that personal actions even within the business arena either do no harm to anyone (including self) or that personal actions contribute to the betterment of the common good. Or purposes may be to always choose to better one's self while attempting to minimize harm to others, a less noble approach. And finally, there is the purpose of human rationalization of each decision in light of the circumstances at hand with the view to justifying any action taken. But hidden within this question of "What is the purpose of ethics in our lives?" are the more basic issues: Why am I here? What am I here to do? What do I really want in this life to make me happy or be at peace? These questions are for each individual to recognize and thoughtfully evaluate. This evaluation process is a very important key element in the progression of personal discovery and in learning about Ethics and Applied Ethics. Without a clear vision of these basic questions, the purposes of ethics, "the noble-leader", "the me-first-good guy" or "the I-got-mine-and-that's-what-it-is-all-about" become muddled, confusing and destructive. When that occurs, we all default to the "I-got-mine-and-that's-what-it-is-all-about" view.

HOW DO WE GET ETHICS?

There is a dispute among scholars that ethics cannot be taught. Clearly, we do not agree with that view. Ethics can and must be taught. There are certain value-based principles that are expected or assumed between human beings depending upon their roles. For example, in the employer-employee role it is assumed that if I hire you to work for me you will not steal from me. The problem occurs when even with a clearly stated policy the employee sometimes silently reserves . . ."unless you catch me." Or, under a contract, parties forget that the agreement was based initially upon good faith. Yet when things go badly during performance the issue become "what does the law say?"

Then there are the questions: Is ethics the same as law? Is it the same as economic decision-making? Is it the same as religious determinations? How, if at all, do these relate in the decision-making process of ethical deliberation? These are questions that clearly show that the subject of ethics must be taught to bring understanding and clarity to the concept and its application. Whether the teaching reinforces previously learned values and methods of decision-making or serves as an introduction to new views on the process of human interrelationships, ethics must be taught. Further, there is a valid view that ethics can best be taught through experience and by the case method as is done in this text. Many other professions, such as engineering, medicine and law, have successfully done so. However, there is a danger in the case method teaching that needs to be addressed. In the case method approach each case is a separate situation. Does this mean that depending upon the situation at hand a person's value-based decision-making may change? We humans are great at rationalizing just about everything, and generally, our rationalization will end up in our favor. Such a method of decision-making may be inconsistent with value principles and, thus, opens the door for others to question our sincerity. It is suggested and proposed that value-based ethics gives us consistent principles for consistent application that not only serve our personal lives and interrelationships but also is a necessary requirement to conduct continuing business operations. Some would say that the law is their value-based principle.

However, I would suggest the law was designed for the purpose of protecting people who have been offended or injured. Law was not designed nor is it intended for proscribing the relationship of people desiring to combine in good faith for a personal or business purpose. Law is there as the last resort when good will and good memories fail or fade. Good will, common view, common purpose is the cement for good business and the common good. Ethics (and applied ethics) using value-based principles is the most consistent way to seek out and assure the common good.

WHAT DO WE DO WITH THE ETHICS THAT WE LEARN AND ACQUIRE?
Think back over your education about the things that you have learned but never used. Remember that ancient history course about the Greek civilization? Remember Socrates, the ancient Greek? Socrates is credited as being the earliest of the questioners and the ethics conscience of his day. He is credited with saying by his disciple Plato, "A wise man knows what he doesn't know." We will always be discovering and learning in our lives. However, there is an implicit understanding that as we learn we also must apply our learning. Ethics is no different. Plato's student, Aristotle, said, "Excellence is not an act but a habit." As we learn to bring values into clear principles for use in our own decision-making process, we are then expected to apply these values in actual decisions. As we do so, we move from a single ethical act to a habit of acting ethically. This is applied ethics. Applied ethics can be defined as a person's consistent and systematic approach to choosing value-based principles for his/her individual standards of conduct and the application of those principles in human relationships.

A KEY CONCEPT TO REMEMBER
Personal ethics, like personal tastes, should not be forced upon others. Ethics, in the beginning and in the end, is a matter of individual choice. This is why education is so important for the discovery, development and application of ethics. One must learn, understand, then choose, and finally lead by example, trusting that other individuals, once enlightened, will choose to join in the never-ending journey of the application of ethics for the common good.

It is obvious to many that "good ethics is good business." When the premise is understood, it heightens our concern for others who are our employees, customers, suppliers, subcontractors, and/or public. Truly showing concern for others in a personal and/or business situation goes a long way in establishing a trusting relationship. That trusting relationship opens the way for all parties to be creative, accommodating and profitable. A non-trusting relationship breeds fear, suspicion, anger, frustration and resorting to the law. It is clear that such negative emotions and relationships create and sustain an environment that is antagonistic to the common good of any personal relationship and/or business arrangement.

Ethical questions seldom arise in a vacuum. This program approaches the issue of ethics in construction by dealing with the business relationships, which exist within the industry. It must be kept in mind that these relationships work both ways. Ethics is a two-way street. Seven typical relationship areas are identified as follows:

- Between Contractors and Owner
- Between Contractors and Architects/Engineers
- Between Contractors and Subcontractors
- Between Contractors and Vendors
- Between Contractors and Other Contractors (firms)
- Between Contractors and Their Employees
- Between Contractors and the General Public (society)

By placing ethical relationships into the above categories it is possible for a program sponsor to concentrate on a specific issue. Thus, the program or class can be made timely (i.e. it addresses an issue of current interest), and it is easier to promote.

It is the hope of the members of the American Institute of Constructors that this publication will make it easy to sponsor a program or class on construction ethics. It is also hoped that the session will generate such positive response among participants that additional programs will be scheduled because of the demand.

ACKNOWLEDGEMENTS

This Introduction was written in support of the American Institute of Constructors and its Program on Construction Ethics by James H. Gill, Jr. JD, Contractors Educational Trust Fund Endowed Chair and Professor of Applied Professional Ethics, at Louisiana State University, College of Engineering, Department of Construction Management, Baton Rouge, Louisiana 70808 (2001).

Case studies were supplied by practicing professional constructors.

CONDUCTING THE PROGRAM

A chapter program on construction ethics can be conducted in about 60 minutes, and is easy to arrange.

SET UP OVERVIEW

1. Select one or possibly two ethical relationships you wish to discuss from the list below:

Relationship	Case Study
A. Contractor - Owner	2, 3, 5, 7, 8, 10, 12, 13, 15, 16, 17, 18, 19, 20, 21, 22, 25, 28, 30, 31, 32, 33, 35, 38
B. Contractor – Architect/Engineer	3, 5, 9, 10, 11, 16, 31
C. Contractor – Subcontractor	1, 4, 7, 11, 19, 23, 24, 29, 30, 34, 35
D. Contractor – Vendor	14, 20, 26, 36, 37
E. Contractor – Contractor	2, 3, 4, 8, 16, 19, 33
F. Contractor – Employee	2, 6, 12, 22, 24, 25, 26, 28, 29, 34, 38
G. Contractor – General Public	6, 7, 13, 27, 31, 38

2. Turn to the cases referenced next to each relationship and select the ones you want to use for a panel discussion. Note that sample questions follow each case. Pick two or three.
3. Select a date, time, and place for the program.
4. Select a moderator and give him or her the cases and questions.
5. Select three or four panelists and give them the cases, **but not the questions**.
6. Promote the program.
7. Format I – Professional Forum Format
    a. Distribute copies of the cases (**but not the questions**) to the audience.

    b. Before starting a session have the moderator briefly summarize the case that will be discussed. The moderator then starts the program by asking the first question.

    c. Audience questions of the panelists are secured via question cards, and can be used by the moderator as he or she sees fit.

STUDENT PROGRAMS AND MODERATOR

Here, student member's act as panelists and audience, a faculty member could serve as the program moderator, and a classroom can easily substitute for a hotel conference center. Student groups should find this a fairly easy program to organize and conduct, and faculty assistance is almost always available. Students should also consider asking local AIC Members to provide office space or serve as moderators.

STEP-BY-STEP

Program administration and promotion: A basic assumption of the American Institute of Constructors is that most if not all AIC chapters know the basics of promoting and sponsoring a seminar, panel discussion, conference, or workshop for members. A PROGRAM ON CONSTRUCTON ETHICS involves the same basic knowledge, and this guideline document does not devote much time to such things as room setups or how soon or often the program should be promoted.

Planning details: The following steps may contain more detail than most chapters need or want. However, they should provide all the information and instructions necessary to conduct a successful program.

1. Fully understand the Objective (Page 2) of the program. Reread it, and be prepared to answer questions about what the program does and is not intended to do.

2. Basic program setups are those for a typical panel presentation with a non-panelists moderator. Figure on a panel (headtable) of three to four individuals with possibly a separate podium for the moderator. The audience can be seated schoolroom, theatre, or in rounds. The only audio-visuals that might be needed are microphones – for the moderator and panel – if the crowd/room warrant.

3. Make a preliminary identification of your target audience. Know that the target may change after further refinement of the issues that will be discussed. However, early identification is a good idea. In most instances the target audience will be construction students, construction contractors, or both. However, as you select the ethical relationship(s) to be featured, a secondary audience may be revealed.

4. Decide on the best time(s) to conduct the program. A PROGRAM ON CONSTRUCTION ETHICS can be held in 60 to 90 minutes, so ideal situations are probably monthly membership meetings and luncheon sessions. Have an alternative in mind.

5. A PROGRAM ON CONSTRUCTION ETHICS can be conducted virtually anywhere – a member's office, a dinner function, a hotel meeting room, a college classroom.

6. Secure support for conducting the program from among the membership. An AIC chapter or student chapter sponsored program on construction ethics should not come as a surprise to key members.

7. Listed below is a series of relationships within which most if not all business dealings transpire in the construction industry. It is here that problems arise. It is here that ethical considerations and questions arise. The program should concentrate on one or more of these relationships, and the choices are yours.

   Select the ethical relation or relationships you wish to cover in the program. For a 60 to 90-minute program select two; for two hours select three. Be advised that the first issue (or case) may generate enough discussion to consume the allocated period. There is nothing per se wrong if this happens. In the same sense, if for some reason an issue or case generates little or no discussion, have a backup case or two ready for the moderator to introduce. All he or she has to do is read it to the panelists and audience.

Ethical Relationships	Cases	Total Cases
Contractors and Owners	2, 3, 5, 7, 8, 10, 12, 13, 15, 16, 17, 18, 19, 20 21, 22, 25, 28, 30, 31, 32, 33, 35, 38	24
Contractors and Architects/Engineers	3, 5, 9, 10, 11, 16, 31	7
Contractors and Subcontractors	1, 4, 7, 11, 19, 23, 24, 30, 34, 35	11
Contractors and Vendors	14, 20, 26, 36, 37	5
Contractors and Other Contractors (Firms)	2, 3, 4, 8, 16, 19, 33	7
Contractors and Their Employees	2, 6, 12, 22, 24, 25, 26, 28, 32, 34, 38	10
Contractors and the General Public (Society)	6, 7, 13, 27, 31, 38	6

8. Once you have identified the relationship(s) that is to be featured in your program, review the cases which apply to the relationship. Each case reveals at least one ethical issue or question that pertains to

the relationship. These are actual (real-life) cases drawn from the experiences of contractor members. Select one or two cases to serve as examples. Choose well, as this selection process is important. These cases will be shared with the moderator, panelists, and program attendees. As noted above in item 7, have backup cases available when you begin the actual program.

NOTE: *If in your opinion none of the cases adequately highlights an ethical question you wish discussed, do not hesitate to modify the case or create your own example. If you are lucky enough to have a member who supplies you with an actual experience (new case), please send a copy to AIC Headquarters for possible inclusion in later editions of this document. Be advised, however, that a new case will require new questions – see Item 10.*

9. Note that on the reverse of every case (or following it) are at least two questions pertaining to ethical issues raised by the case. Study them. Feel free to modify the questions or add others if you think changes will better address an issue you want discussed. These questions are for use by the program moderator, not the audience or panelists.

10. Select potential panelists for the program. In securing panelists fully explain the objective of the program – what it is intended to do and not do. Tell them they will be given a copy of the case(s) well in advance of the program, and discuss the types of questions that they will be asked to address by the moderator. Note that no other cases (examples) will be used – no surprises – without their permission. However, point out that audience questions via question cards will be solicited. Tell them that the moderator will make every attempt to eliminate any "bad" questions should they arise, and panelists are under no obligation to respond to anything they feel is non-germane. Also tell them, if possible, who else will be part of the panel. As follow-up, be sure to mail the cases (not the related questions) promptly to panelists – well in advance of the program.

    This way, if someone is reluctant to discuss a certain case (issue), there is time to secure a substitute case or panelist. Remember, panelists must be volunteers only.

    In selecting panelists, take a good look at the relationship that is being featured and the case or cases that serve as an example. It may (or may not) be quite appropriate to add a non-contractor to the panel. The addition of an architect to a panel discussing ethical relations between contractors and architects might prove interesting, same for a subcontractor when discussing that relationship, and so forth. However, what about adding an attorney to a panel with contractors and vendors? A public works official to a panel with contractors and engineers? Give the matter some thought.

11. Secure a moderator. He or she does not necessarily have to be a contractor member, but should certainly be familiar with the industry. If you want, you can select the moderator before you select the panelists, and you can involve the moderator in deciding which cases will be discussed. Student chapter sponsors can avail themselves of fellow members and construction program faculty as moderators, but as mentioned earlier, a local contractor might welcome this role.

12. After the ethical relationship(s) are identified and the moderator and panelists selected, it is time to promote the program – if not already started. Promotion is mentioned here because the final selection of the relationship and panel may have altered your preliminary identification of the audience. It is possible that in addition to contractor members, the program can be promoted to other groups such as vendors and subcontractors.

13. When the audience arrives for the program, have two things ready for them (also see item 15): a copy of the case(s) to be discussed (without the related questions on the reverse), and a participant question card or sheet. Distribution of the cases is extremely important, and the audience must be given time to read the appropriate case just before the panel presentation begins. It truly acts to involve them in the

discussions. Obviously, the time needed will depend on the length of the case. A sample question card appears at the end of this section.

14. The moderator should begin the panel discussion with a brief review of the case. Since all will have read it, no more than a few sentences are needed. The moderator then begins the panel discussion. He or she can start with a specific question addressed to a specific panelist, or with a general question – to whichever one decides to answer first. Much will depend on the case being discussed. The questions appearing on the reverse of or following each case in this guideline document are for the moderator's possible use.

15. Since general tips for moderators include asking open-ended questions such as, "Bill, how do you view this issue with respect to your firm," and asking questions that call for opinions, such as, "do you agree with John's comments, and if so why – or why not?" Follow up questions to something a panelist has said are important.

16. Dealing with question cards from the audience has both good and bad features. It certainly involves more people (the audience) in discussions, and audience questions are frequently timely and excellent – bringing out points and issues not otherwise discussed. On the other side is the fact that the questions must be collected, (requires another person), and the moderator should look them over before giving them (reading them) to the panel or a specific panelist. Duplicate questions should be combined: questions that could conceivably lead to agreements that illegally restrain competition should be ignored. If a large number of questions are received, the moderator must also know when to cut them off, and move on to another case.

    It is far simpler to eliminate question cards and simply entertain questions from the floor. Unfortunately, this affords little control over what will be asked. This may not be an issue for student chapters, but it can lead to an occasional awkward situation among a panel and audience of all practitioners.

    The choice of methods in securing audience questions for the panel is up to the moderator and panelists. Ask for their thoughts on the matter. Regardless of the method, some form of audience questioning should be provided. When you ask for it; how will you get it? It's up to you.

1. The moderator should thank the panelists at the conclusion, lead a round of applause for them, and end the program on time – or close to it.

VARIATIONS ON THEME

IF A PROGRAM ON CONSTRUCTION ETHICS is to be conducted at a dinner meeting, just as dinner starts the moderator charges the audience with the following task.

a) They are to read a case that was given them (or is in their chair) when they arrived. This can be a special case, one that is not going to be covered in the panel discussion, or it might be one of the panel discussion cases.
b) They are to come to a consensus among those at the table concerning a question or action revealed by the case. (The moderator must select the question/issue and read it aloud to the audience at least twice.)
c) They are to appoint a spokesperson for the table who will report their consensus after the panel presentation.
d) Provide enough time, after the panel discussion program, for each table spokesperson to report – perhaps 60 seconds per table.

AUDIENCE QUESTION CARD

Please write or print clearly, and be as brief as possible.

Question: _____

_____

_____

_____

_____

_____

This question is directed to (optional): _____

AMERICAN INSTITUTE OF CONSTRUCTORS CODE OF ETHICS FOR CONSTRUCTORS

The construction profession relies upon a system of ethical competence, management excellence, and fair dealing in undertaking complex works to serve the public with safety, efficiency, and economy. The members of the American Institute of Constructors are committed to the following STANDARDS OF PROFESSIONAL CONDUCT:

I. A member shall have full regard to the public interest in fulfilling his or her responsibilities to the employer or client.

II. A member shall not engage in any deceptive practice, or in any practice which creates an unfair advantage for the member or another.

III. A member shall not maliciously or recklessly injure or attempt to injure, whether directly or indirectly, the professional reputation of others.

IV. A member shall ensure that when providing a service which includes advice, such advice shall be fair and unbiased.

V. A member shall not divulge to any person, firm or company, information of a confidential nature acquired during the course of professional activities.

VI. A member shall carry out responsibilities in accordance with current professional practice, so far as it lies within his or her power.

VII. A member shall keep informed of new thought and development in the construction process appropriate to the type and level of his or her responsibilities, and shall support research and the educational processes associated with the construction profession.

Standard of Conduct for Salaried Employees: Pizzagalli Construction Company

The Company (including for these purposes, all of its affiliated entities), relies on the integrity and good judgment of all employees to observe ethical, professional and legal codes and good business practice in the conduct of Company affairs. It is the employee's responsibility to avoid any arrangement, agreement, investment, employment relationship, act or interest which actually is, or appears to be, contrary to the best interests of the Company or its clients.

It is not possible to delineate in the context of this policy statement all of the possible situations which might constitute conflicts of interest or unethical conduct, and employees are expected to be guided by the general standards set forth herein. In any situation where the employee has any doubt whatsoever as to whether a given course of conduct might be contrary to such standards, the employee should discuss the matter with the Company's senior management in advance of involvement or participation.

During the course of employment with the Company, employees may come in contact with confidential or proprietary information concerning the Company's activities. Employees shall not disclose such knowledge, acquired because of the position of trust held, will not be disclosed to others except as specifically authorized or required by the Company.

The Company believes that every staff member should devote his/her full time and ability during regular hours of employment to the services of the Company, and , when conditions require, will not allow the Company's interests to be neglected outside of such regular working hours.

Generally, there is no objection to any employee's involvement in a business investment where (1) the employee is not required to devote any time to the investment during regular work hours, (2) the investment is not an enterprise which in any way competes with the Company, (3) there is no significant business relationship between the enterprise and the Company, (4) the investment does not require the utilization of any unique or particular skills, experience or knowledge gained by the employee while in the employ of the Company,

(5) the investment will not be perceived by the public as being connected in any way with the Company.

Examples of unethical practices include the giving or accepting of monetary, material, or other consideration as a condition for effecting a transaction, the imposition of unusual or unreasonable requirements and/or restrictions which tend to limit competition and favor a single party, improperly dealing through or with a person or position of influence to gain an advantage for the Company and releasing without authorization confidential or proprietary information.

No employee of the Company may solicit, encourage or accept any gift, rebate or other form of consideration from any vendor or subcontractor, whether at Christmas or any other time throughout the year. It is a Company policy that all purchasing be done on the basis of price, quality, performance and other pertinent factors in the best interests of the Company. The Company will, however, make an exception for small Christmas gifts of a nominal value up to $10; however, it encourages our subcontractors and vendors to make no gifts at all.

The basic principles underlying the above standards of conduct preclude an employee's use of members of his/her family as a device or subterfuge to circumvent any of these policies.

(Reprinted with permission from Pizzagalli Construction Company, Burlington, Vermont.)

HCB Contractors: A Reputation of Integrity

HCB Contractors and its employees have earned an outstanding reputation for integrity and good citizenship. The well being of both the Company and its employees depends greatly upon the continuance of our excellent reputation. However, no matter how fine a reputation a company or its employees enjoy, that reputation can be seriously damaged by improper conduct.

Although it is not the intent to impose any form of regulation on the personal and private affairs of any employee, it is the policy of HCB Contractors that employees meet certain standards of business conduct. Employee conduct, both on and off the job, reflects upon the character of the Company; therefore adherence to these standards by everyone is the only sure way HDB Contractors can continue to merit the confidence and support of our clients and the communities in which we work.

This policy is general in nature and does not include all the rules and regulations that apply to every situation. The policy sets only the minimum performance standards and must be viewed within the framework of Company policies, practices, and instructions.

Personal standards of performance:

1. All employees will maintain congenial, yet professional, relationships with clients, architects, engineers, subcontractors, and suppliers of goods and services to the Company. No employee shall have any financial or other relationship with business contacts that might impair, or even appear to impair independence of judgment on behalf of the Company. In other words, action must be avoided that involve, or appear to involve, "conflict of interest" – both in business relationships and in personal activities in the community.

    Employees must not use HCB labor or influence others to have work performed on their personal home or personal investment property.

    Employees must not accept cash for any reason. Favors, gifts, trips, entertainment, or anything of substance (i.e. over $50) from vendors, suppliers, clients, subcontractors or others with whom the Company does business must be reported to the employees' respective District Manager or Vice President.

    Loans must not be accepted from any persons or companies having or seeking business with the Company, except recognized financial institutions.

    Fully stated, our policy is to award business based on merit, without favoritism, to qualified, responsible bidders at the lowest price.

2. At all times, employees should give full attention to meeting their assigned duties and responsibilities, keeping personal business affairs separate and distinct from Company business.

    No employee should accept employment or engage in a business activity which involves:
    a. Any activity during hours of employment with the Company;
    b. The use of any Company equipment, supplies, or property;
    c. Any direct relationship with the Company business or operations.

3. Doing business with the firms in which employees have a direct or indirect financial interest is not permitted unless fully disclosed and with concurrence of the President of the Company.

4. No employee of the Company shall directly or indirectly use or repeat for personal gain any information acquired as a result of employment with the Company.

    Company business records must be prepared accurately and reliably. They are of critical importance in meeting our financial and legal obligations.

    All reports, vouchers, bills, payroll and service records, measurement and performance records and other essential data must be prepared with care and integrity. Records are to be carefully safeguarded and kept current, relevant and accurate. They are confidential and to be disclosed only to management personnel and others authorized to receive this information.

5. Each person is personally accountable for funds over which he or she has control.

    Anyone spending money that will be reimbursed should always be sure the Company receives good value in return.

    Anyone approving or certifying the correctness of a voucher or invoices should have the knowledge that the purchases and amounts are proper and accurate.

Anyone responsible for the handling of Company funds or revenues, as well as associated records and materials, will be held accountable for the safekeeping of these items.

6. Employees are required to provide the Company's internal and outside auditors with any and all information necessary for the performance of their responsibilities. Concealment of financial and operating information, inaccurate cost reporting, or providing misleading information is a violation of Company policy.

7. Protection of Company property is the responsibility of every employee. It is imperative that we prevent fraudulent or negligent misuse or theft of our property.

   Company property may not be used for personal benefit. It cannot be sold, loaned, given away or otherwise disposed of, regardless of condition or value, except with proper authorization by an official of the Company.

8. HCB Contractors does not contribute financial or other support to political candidates, organizations, or parties. No employee is authorized to make or approve any contribution on behalf of the Company. However, the Company encourages its employees to participate in the political process through personal support of the candidate or political party of their choice. Any pressure, direct or implied, that infringes upon the right of any employee to decide whether, to whom, and in what amount he or she will make a political contribution is strictly forbidden.

9. Integrity is a personal responsibility. Each employee assumes a position of trust, and is accountable for his or her actions. Violations of these trust subjects one to disciplinary action including dismissal and any other remedies available to the Company.

(Reprinted with permission from HCB Contractors, Dallas, Texas.)

CASE STUDY NO. 1

As a general contractor, Conway Construction Corporation entered into written contract with Setcher Roofing Company for all roofing, decking, and sheet metal on the Bentonville High School for a contract amount of $261,000.

This was one of several contracts that CCC had with Setcher and was one of many contracts over a period of years prior to this contract. During the course of the job there were three change orders for additional work; one for $1,060; one for $2,380 and one for $21,540. All change orders were approved by the architects and paid by the owners. During the course of the job, CCC received and paid several monthly estimates which included the base contract on two of the smaller change orders for some reason unknown to CCC, Setcher failed to acknowledge the $21,540 change order or invoice CCC for same. The work covered in the change order was, however, completed on schedule as per plans and specs.

Some five months later after an internal audit by CCC, it was called to top management's attention that Setcher had signed all lien releases and gave CCC a paid in full receipt back some four months ago.

The ethical question asked was, do we keep quiet and hope Setcher & Company never finds out that they were not paid for the $21,540 change order, or do we pay as per our contract provisions?

Relationships and Questions

Relationships Featured

Contractor to Subcontractor

Questions

1. Is there really any ethical problem here?

2. What if the Setcher firm was not an old associate, but a firm with a terrible reputation for cheating others and you expected never to work with them again? Is your decision different? Why?

CASE STUDY NO. 2

Tom O'Kay was a large contractor doing major projects for the Fortune 500 type corporations anywhere in the country. His types of clients were hard to get, but once you had proven yourself, there were continuous opportunities for work. These huge corporations had a steady stream of large projects in various locations that were generally bid out of a corporate headquarters. One particular client that Tom had not been able to get work from had two major projects in his state. A contractor that Tom knew to be a fine contractor from another area, did most of the work for this client, and had built the two projects in Tom's state.

Tom learned that the corporation was ready to double the size of one of the original projects in the state. Because of the success of the first two projects, and the extremely tight schedule dictated by the company's needs, they had decided to award it to the original contractor on a CM contract without competition.

Shortly before the project was to start, Tom heard a rumor that the competing contractor was having financial difficulty. One of Tom's employees used to work for the contractor, and Tom had him call an old friend at the firm and make some inquiries. It turned out that the contractor was indeed having some difficulties, and more particularly, the firm could get no new bonds pending a review of its current situation.

With this information in hand, Tom immediately called the potential client and got through to a senior vice president. Tom stated that he had heard some serious information concerning the project in Tom's state. Tom said he felt he was duty bound to advise the company of the other contractor's difficulties. After several phone calls over as many days the senior officer informed Tom that they had confronted the contractor, and he had assured them that the project would be handled in the same manner as the others, and that the rumors about his difficulties were greatly exaggerated. Tom, frustrated by the answer, persisted. He advised that the corporation, to protect itself, should require a bond, at the very least. The company wanted to use the contractor who had completed the original building, but was concerned enough to request the bond. It could not be produced.

As the schedule was now more of a problem than ever because of the time lost due to these complications, Tom's company was selected to manage the project.

Relationships and Questions

Relationships Featured

Contractor to Owner
Contractor to Other Contractors
Contractor Employer to Contractor Employees

Questions

1. Was Tom acting appropriately when he asked his employee to contact friends at his old employer's office and secure financial information?

2. Were Tom's actions in this matter (advising the owner of his competitor's difficulties) within the bounds of fair competition, and is fair the same as ethical? Would it make a difference if the competitor was, in fact, known for his poor workmanship and unfair treatment of subs and suppliers?

3. If Tom did not tell the owner of his knowledge that his competitor was having problems, would that be proper?

4. Does a contractor, such as Tom's competitor, have an obligation to advise the owner of his financial difficulties?

CASE STUDY NO. 3

In March of last year I was paid a visit from the pastor of a church in a nearby community who was trying to build a 41 unit elderly housing project.

He had been dealing with the state's Housing and Economic Development Authority regarding inexpensive financing, and had heard that we recently placed into operation a very successful 40 unit elderly housing project using conventional financing. The pastor was obviously interested in what we had done, and impressed with how we did it.

After lengthy discussion we shared with him our entire experience on the project, and went on to do some estimating work which would have ordinarily been done only under a contract. We completed this work because he seemed genuine and I felt we had a very good chance to land the construction management contract because of the similarity between theirs and our projects. Besides, as a matter of course we always do a little extra for a church organization.

After several more weeks and more estimating work we requested a meeting with their building board in the hope of finally securing a contract. They invited us to a meeting. By this time the architect, unbeknownst to us, had involved a few competitors who were also to be interviewed the same day.

I had my accountant there with me for the presentation as their main man who seemed to have the most to say was an accountant, not the building chairman, and liked to talk to accounting people. I was informed of this prior to the meeting by the pastor.

Upon arriving, the building chairman was introduced to us by the pastor and was ushered into the meeting room where about 20 people were seated. The pastor quickly reviewed his dealings with us, and by so doing introduced our company. We were not afforded the same courtesy of introductions from their side, and perhaps I should have asked.

The interview was carried on as most contractor/church council meetings might be: the usual presentation, the usual questions, and then the pressing for hard details, all of which you can

expect. However, some of the questions concerning percentage fees and building procedures were far more detailed than normal. By this time I had figured out who the architect was. It almost seemed like he was trying to conceal his identity.

After we were ushered from the room my accountant blurted out, "did you realize that a competitor of yours was in the meeting?" I replied that I did not.

The next day I called the un-introduced architect and protested that our interview was unfair and should be re-evaluated. I told him he should have had the ethical courtesy to stop that meeting and inform me that a competitor was in the room, or he should not have allowed the man in the room in the first place. He said he felt it was OK for the competitor to be there because his presentation was so "different" from mine, that no impropriety existed in his eyes.

I later spoke to the pastor who was very embarrassed by the whole matter. It was obvious that he was not pleased, but felt that the project had come so far with the architect they could not turn back.

The end result is that we did not get the job. It was awarded to a third contractor, a friend of the architect who made the last presentation at the meeting. He was also in the audience during our meeting.

Relationships and Questions

Relationships Featured

Contractor to Owner
Contractor to Architect/Engineer
Contractor to Other Contractors

Questions

1. Is this a demonstration of good professional behavior by the architect? What or how much leeway does an architect have?

2. Can the architect's actions be justified if he is, in fact, making an honest attempt to secure the best deal for his client?

3. What, if any, are the owner's ethical obligations to the contractors who were involved in this case?

4. Did the contractors seated in the audience have any obligation to identify themselves? Would your answer be different if you were asked by the owner or architect to sit in on a competitor's presentation?

CASE STUDY NO. 4

In South Dakota, the contractors have set an informal deadline for materials and subcontractors' quotations prior to State Highway lettings. The lettings are at 10:00 AM, and the informal deadline is 12:00 midnight, just prior to the letting.

At a recent letting we had several grading sub-bids by midnight, the lowest of which was Rossard. At 5:00 AM we received a call from Standard Paving, and took a quotation from them on grading. It proved to be $2,000 lower than Rossard. We were low bidder on the project and gave the grading to Standard Paving. We feel we owed Standard Paving the job because we accepted their 5:00 AM bid – and they were low. Rossard was understandably upset.

Relationships and Questions

Relationships Featured

Contractor to Subcontractor
Contractor to Other Contractors

Questions

1. Was the contractor ethically correct in taking and/or using the bid from Standard Paving?

2. What if the Rossard Paving sub-bid was used in your bid, but you then gave the job to Standard?

3. Is the sealed bid process undermined by informal agreements or deadlines for materials and subcontract quotations?

4. Was the contractor wrong in accepting Standard's bid after the informal deadline?

CASE STUDY NO. 5

A contractor was performing the construction of a 100,000 square foot concrete frame building. The owner was doing the site inspection and referring questions to the Architect/Engineer for resolution.

After placing the south half (15,000 sf) of the second of the three floors, a concern was raised by the general contractor's superintendent regarding the yield of the concrete on the pour. Field investigation by the contractor's staff identified the problem as an error in set-up of a laser that caused the slab to be poured one-half inch too thick. (There were no embedments or edge conditions to indicate the extra thickness.)

The contractor identified the following options:

1. Do not expose the error. Pour the remaining half of the slab one-half inch thicker than the design, and adjust the other materials to fit.
2. Do not expose the error. Pour the remaining half of the slab as detailed, and feather to the thicker slab in the adjacent bay.
3. Expose the problem. Verify that the extra dead load imposed by the additional one-half inch of concrete does not pose a structural hazard, and seek the Owner's and Architect/Engineer's approval to either thicken the remaining slab or feather out the difference in elevation.
4. Remove the entire slab and reconstruct it at the specified thickness.
5. Find a way to remove the one-half inch.

At a meeting of the contractor's management it was decided to pursue option Number 3 for the following reasons:

1. It is the contractor's policy not to hide variances from the job specifications.
2. The structural implications seemed minor and there appeared to be several reasonable options to solve the architectural detailing problems.
3. Whatever remedial work was required would be much less expensive prior to preceeding with construction.

The problem, along with the three proposed methods for solution, was presented to the owner who referred it to the Architect/Engineer. While agreeing that the structural implications were minor and that the detailing could be revised, the Architect/Engineer's position was that the owner should not accept any error by the contractor, and that the contractor should be instructed to provide a slab that met the specified tolerance. The owner so instructed the contractor.

After several meetings and a round of letters to protest the unreasonableness of this demand, the contractor concluded that he must either perform options 4 or 5, or become involved in a lengthy dispute that would seriously damage the project schedule.

Concrete planning equipment was brought in and the one-half inch was removed and the slabs refinished at a cost of $25,000.

In hindsight, should the contractor have changed his actions? The contractor believes he did the proper thing with regard to the error, however it was agreed among the contractor's staff that in the future greater effort would be expended assessing whether the other parties are likely to be reasonable prior to entering into a contract.

A footnote:

While it is the contractor's policy not to hide any error, it is not his policy to unilaterally pay for unreasonable actions by others. Therefore, when the opportunity arose to price some changes to the work on this project, the proposal price was adjusted prior to submission in order to cover the addition cost of the slab work.

Relationships and Questions

Relationships Featured

Contractor to Owner
Contractor to Architect/Engineer

Questions

1. Is this really an ethical problem or simply a technical problem?

2. Should the contractor have disclosed the error? Why or why not?

3. Does the owner have a reasonable right to expect and demand that the contractor's work will be done in exact accordance with the plans and specifications?

4. Does a contractor have a right to expect and demand that architects and owners apply reason and good judgment in measuring the standards of his work?

5. Is the contractor justified in "getting even" with the architect by "overpricing" change orders?

CASE STUDY NO. 6

The project I'm managing has not been going well. There was, first, an old but working "sewer" where none was supposed to be. It was probably for water drainage, buried when a swamp was filled in to make dry land for faring or settlement purposes a hundred years ago. By the time we had repaired the sewer and relocated the building's site, we had lost two weeks. Now the trenchers have unearthed 19th Century gravestones, six so far, it suggests that the site might be a forgotten graveyard. If it is a graveyard, we have to stop all work until the graves are moved. Opening a grave without legal authority is a crime.

When I phoned my boss with the bad news, he just said "Don't worry. Did you ever hear of a graveyard in a swamp? The gravestones are probably just part of the fill. Ignore them. Don't tell anybody. We can't afford to lose any more time."

My boss is right, of course. We can't afford to spend a week or two waiting for someone from the county to make up his mind about whether the gravestones are just debris, or, instead, mark a buried graveyard. On the other hand, though, I'm the one in charge of the job. If this is a graveyard, and I don't notify the authorities, I'm the one who will be responsible for the desecration.

Relationships and Questions

Relationships Featured

Contractor to Contractor Employees
Contractor to General Public (Society)

Questions

1. Is this an ethical question or legal question?

2. What should the project manager do and why?

3. What would you do if your boss told you to do something which you knew was unethical?

4. If a person pretends they do not know something, or just avoids telling what they do know, is it the same as lying?

CASE STUDY NO. 7

A few years ago we were low bidder on a complete remodel of the Big Lake Courthouse in Cheyenne Falls. The lowest mechanical sub bid we had was from a firm we had not solicited, and who called in their price to our secretary about 30 minutes before the letting. We had a previous unsatisfactory experience with this mechanical sub-contractor, and elected to bid the project using the price of the second lowest bidder – who we had solicited. When we didn't offer the job to the low bidder, he protested to the County Commission that we had divulged his price to the second bidder and that we were treating him unfairly. Of course the County Commission took no action in the matter, leaving subcontractor selection up to us.

Relationships and Questions

Relationships Featured

Contractor to Owner
Contractor to Subcontractor
Contractor to General Public (Society)

Questions

1. If a public agency is required to award construction contracts to the lowest bidder, shouldn't the same rule apply to contractors with respect to sub contracts?

2. What is a contractor's obligation to an unsolicited bid?

3. What if the contractor had, regardless of past experiences, solicited the bid?

CASE STUDY NO. 8

A local developer advertised for bids for site improvements on a large project. Four bidders submitted quotations for the project. The bid opening revealed that bidder A was $25,000 below bidders B, C, D. and were substantially higher.

Following the bid opening bidder B claimed an error in their bid and asked the developer to be allowed to correct the mistake by adjusting their price downward by $30,000.

The effect of this would be to make bidder B, the low bidder. Claiming the right to waive any inconsistencies the developers accepted bidder B's revised bid, and awarded the project to him.

Relationships and Questions

Relationships Featured

Contractor to Owner
Contractor to Other Contractors

Questions

1. Was the developer's decision the right one? Why or why not?

2. Would your opinion about the developer's actions change if bidder B did not know what bidder A price was prior to changing his bid?

3. Is the bidder right in offering a lower bid after all other bids have been divulged?

CASE STUDY NO. 9

The architect on a masonry project in the northern part of the country insisted that a minimum temperature of 50 degrees be maintained when constructing the masonry. The contractor erected shelters and provided temporary heat at or above the 50 degree level, and the architect performed several spot checks to insure compliance.

One night part of the system failed. The architect checked the temperature at 5:00 PM and found it to be 52 degrees. A check by the architect at 10:00 PM found the temperature to be 37 degrees. A further check at 5:00 AM showed the temperature to be 30 degrees. At 7:00 AM the contractor was notified that the project was shut down.

The architect made no attempt to contact the contractor even though he knew at 10:00 PM the previous night that there was a problem with the system. Claiming that the mortar had frozen, the architect rejected a substantial portion of the installed brickwork. Although independent lab tests showed that the mortar surpassed design strength, the contractor had to replace all the brickwork in dispute.

Relationships and Questions

Relationships Featured

Contractor to Architect/Engineer

Questions

1. Is this really an ethical issue or simply a technical dispute?

2. Was the architect under any ethical obligation to advise the contractor that the heating system had failed?

CASE STUDY NO. 10

A room was not listed on a room finish schedule for a remodeling project. After bid and award of the contract, the architect selected colors for the carpet, base, wall treatment, and ceiling of the unscheduled room.

The contractor objected, and was given work (change) orders by the architect for the work. However, the owner refused to pay the additional costs for the architect's error.

The contractor soon found that payment requests were processed slower and slower. At job close out, vast amounts of money were withheld for minor items. Upon correction of the items a new list of minor items appeared. In desperation, the contractor dropped his claims for payment against the owner, and the building was accepted.

Relationships and Questions

Relationships Featured

Contractor to Owner
Contractor to Architect/Engineer

Questions

1. Is this a form of extortion?

2. If you were the contractor in this situation, what would you do and why?

3. Was it ethical for the owner to withhold payment after the architect had approved the change order?

4. Do you feel that the owner has any obligation to support the contractor's position in this case?

5. If the contractor knew that the room was left out of the finish schedule before his bid, did he have any ethical duty to advise the architect and/or owner of this error?

6. Was the contractor's request for a change order on this room ethical when the room was shown on the plans even though left off the finish schedule?

CASE STUDY NO. 11

After months of effort, we recently made the short list of a major private owner, and were asked to submit a bid for one of their stores in the Northwest. We were told to list our major subs and their bid amounts in our offer. We did as requested.

A few days after submission the owner's representative called and advised that we had been selected do to the work. A meeting was arranged for that afternoon. At the meeting we were offered the contract, but at a lower amount than the bid price. Along with the contract was a list of sub bids with lower prices than the ones we had used to compute the contract amount. The owner's representative advised that he had combined the lowest sub prices from all contractor bids to arrive at the price, and we could use these prices to reduce our costs – or not sign the contract.

I signed the contract, returned to the office and began calling subs – both ours and the ones on the list.

Relationships and Questions

Relationships Featured

Contractor to Architect/Engineer
Contractor to Subcontractor

Questions

1. Has the owner's representative violated any rules by selecting the lowest sub bids to arrive at his final price?

2. Should a contractor have to accept a contract under these conditions? Would you? Why or why not?

3. Under such conditions, is the contractor acting in good faith by contracting subs who did not originally submit prices to him?

4. How fair is this to the subs who the low bidder used in his original proposal?

CASE STUDY NO. 12

It was late June and very hot. I had three fence crews working the right-of-way of a Virginia interstate, and we were in an area where every bottom turned out to be a mini-swamp. The specs called for three sizes of fence posts: 4" x 4" x 8', 4" x 4" x 12', and 10" x 10" x 8', the smallest being the typical post, with the others used every so many feet to add strength for corners and/or pulling wire.

The swamps were so full of snakes that over half of my men refused to go into them. To add to our problems, the standard 8' posts wouldn't work as there was nothing but water and muck to hold them down. We could usually get the 12' post into something solid, but the plans did not call for their use in most instances. We used them anyway when we had to – and the state inspector on the site either didn't notice or didn't care. Then we came to a solid, 100-yard, no question about it swamp.

I discussed the potential problems with the project manager. He said that I should use 12' posts only if we absolutely had to and could get away with it – otherwise, "make it look good, even if you have to prop the 8' posts up with mud and sticks." He refused to request a change order from the state because (1) he was already having more significant problems with inspectors and change orders on another portion of the job, (2) there was not much money in the fencing anyway, and we already had the needed posts in stock, (3) a change order would take too much time to process, and (4) the state would probably end up requiring even larger 12" x 12" posts which would take us two weeks to get and a lot longer to put in. He was correct on all counts, and I especially didn't want to be lugging even larger 12' posts through a swamp.

What he didn't mention, however, was if I did what he said and got caught, we would be back in here in late July or August doing the work again! It might cost me a crew.

I went back to the job and discussed the matter with my swamper foreman. He suggested that we get together with the state inspector assigned to the work, a move that surprised me. I soon found out why.

It seems the inspector was fully aware of our problems. He knew that my men had been propping some posts up with sticks and using 12' posts where they were not called for. He said that using the longer posts was fine by him as they exceeded specs. Nevertheless, their use should have been reported and a change order requested – a paperwork hassle he didn't want to mess with either. He said he'd been looking the other way because he wanted out of the swamps as much as us. The guy then produced a field book indicating every place we'd switched posts or cut a corner. There were a lot more corners cut than I knew about, even missing posts, and my foreman owned up to the facts on the spot. He said he simply hadn't told me of every instance a problem because there was probably nothing I could do about it. He was right.

The inspector offered to cut a deal. He said if we fixed two places where it really looked bad; used the larger posts where he said – which would be where he thought they were truly needed regardless of the plans; made the rest "look good" with mud or sticks or whatever; and did the remaining dry ground work according to the plans, he would continue to go on appropriate coffee breaks – if I bought the coffee. However, if we did not do exactly what he wanted he would insist on following the letter of the plans (which would force a change order request with resulting delays and even bigger posts). He then said his job was now on the line, but if I reported him to his boss or my project manager he would hand over his book to the state and we'd end up doing and redoing a lot more work than he was asking.

To do what the inspector wanted would call for us to eventually use more of the larger posts than the project manager wanted or we had in stock, a fact I could cover up for only so long. On the other hand, I could always claim a shipping error must have occurred. I didn't know what to do – especially in light of the corner cutting my best and only swamper foreman had not told me about. The inspector and I shook hands.

Relationships and Questions

Relationships Featured

Contractor to Owner
Contractor to Contractor Employees

Questions

1. Is there anything wrong with the deal struck between the superintendent and inspector?

2. What are the superintendent's obligations to inform his employer of the inspector's offer?

3. Is the owner's representative (the inspector) dealing fairly with the superintendent? Why or why not?

4. Is the superintendent's decision to lie to his project manager (if necessary) about the excessive use of larger posts an appropriate one?

5. Should the foreman be disciplined for not telling his boss the entire truth, or was he justified in withholding some information?

6. If you were the fencing superintendent would you have had any concerns over following the project manager's directions? Why or why not?

7. Everyone is this case seems to resist doing a change order for the fence work. Could this indicate that the origin of the problem here might be something other than the swamps?

CASE STUDY NO. 13

On a small highway job in the foothills of western Virginia, a call came over the radio from one of my foremen. A woman had taken a few shots at our fence crew and run them off the job. By the time I arrived at the scene the state police, six state inspectors, my superintendent, and the fencing foreman were waiting. Two of the field inspectors had already gone up the hill to resolve the problem with the property owner and had also been shot at – by a man. After much discussion it was decided that the state's Resident Engineer who had just arrived and I would accompany (follow) the state trooper up to see if we could resolve the problem.

At the top of the hill, just across the right-of-way, was a very small, old frame house. As we approached, a man with a rifle and a woman carrying an infant and a shotgun stepped outside and told us to stop. The trooper told them we wanted to talk, and they lowered the weapons. The man named Gibson, told us that the right-of-way fence was on his property and, in fact, was about to separate his home from his well. He pointed, and sure enough the well was inside the right of way. The state engineer got out the plans and concluded that our fence was located properly. He told the Gibson they had been notified of the project long ago, and the state had paid them for the land. "Why didn't you say something two years ago?"

The engineer said that since drinking water was involved Gibson had a perfect right to get an attorney and appeal or protest the state boundary. The appeal would probably win a change in the right of way if this was his only source of water. The Chief Inspector, who was still angry about having his men shot at, had joined us by then and added that while the protest would surely be considered, the fence, meanwhile, was going to go in – even if he had to have state troopers stationed at the work site. He said Gibson had been paid for it; had accepted the money; and the law is the law; and that we have the legal right to build the road. The only thing the man said, and very softly, was that we could do what we wanted, and that he didn't need a lawyer or judge to tell him what was his property or his father's or grandfather's well.

We started to walk away and I got an idea. I suggested that I could change the schedule and put my men to work somewhere else for at least two weeks or so while Gibson's protest was being processed. (It was a lie. I had little else for them to do.) This would also allow time for everybody to cool down, I added. The inspector agreed, the trooper looked relieved, and Mr. Gibson said we could do whatever we want. We all left.

It was my feeling that the man would not cooperate in any formal protest procedure, would never get an attorney, probably couldn't return any money he'd received, and that he would surely shoot someone. I could also not afford the two week wait I'd agreed to. That afternoon I drove to the nearest hardware store and purchased a new chain saw, work gloves, a shovel, and two or three different size wire cutters. That evening I drove to the Gibson place and had an honest talk with him. I told him that if he would let me put the fence in tomorrow, and leave it for one week, he could tear it out any time he wanted after that. I told him that the chances were good that the state would not be back to check for at least 10 years. The tools were his to use and keep if he agreed. He did.

The next morning I got my fence crew and a few others to put the fence in immediately. I called the Chief Inspector and told him that I had spoken to Gibson and that he'd changed his mind. He wouldn't bother us or protest anything if we got the fence installed in one day and never came back. "Do it at once, one inspection, and get out!" The inspector agreed immediately, cleared it with his superiors, and came out himself that afternoon to perform a final inspection of the work. I knew he'd agree to anything that would reduce his paperwork and eliminate a problem – making him look good.

I don't think anyone's been back in ten years, and I don't think I did one thing wrong. My foreman, however, still says I bribed the guy out of asking for the return of his rightful property. He says that Gibson probably can't read or write, and I took advantage of his ignorance. I feel I did everybody a real favor.

Relationships and Questions

Relationships Featured

Contractor to Owner
Contractor to General Public (Society)

Questions

1. By his actions did the contractor assist the owner (state) in unjustly "taking" property away from a private citizen?

2. Do you think the contractor took advantage of the ignorance of the property owner in order to complete his work?

3. Do you feel the contractor had a right to circumvent the owner's (state) due process procedures for correcting a boundary dispute with a land owner? Did he take the law into his own hand?

4. Was the gift of the chain saw and the other items a bribe? Did the end justify the means here?

5. Was every action taken ethical? If not, why not?

CASE STUDY NO. 14

A building contractor has an agreement with a manufacturer of a pre-engineered metal building system and uses the manufacture's products on both competitive bid jobs and on negotiated jobs. When the contractor is faced with a competitive bid situation, the pre-engineered metal building manufacturer offers the contractor a "strategic discount" to make the contractor more competitive. When the same contractor uses the pre-engineered metal building manufactures products on a project that is with a regular client and price competition in not a factor, the pre-engineered metal building manufacturer does not offer the "strategic discount." In essence, this means that the clients pay more for the product than the contractor's one-time or sporadic clients.

Relationships and Questions

Relationships Featured

Contractor to Owner
Contractor to Vender

Questions

1. Should the contractor continue to accept strategic discounts on competitively bid projects and allow his regular clients to pay more?

2. Should the contactor ask that the pre-engineered metal building manufacturer offer the same type discount to a regular client on a non-competitive bid situation and if they refuse then...

3. Should the contractor, in order to be fair to his regular clients, tell the pre-engineered metal building manufacturer that the project in question is to be competitively bid (which would be a falsehood)?

CASE STUDY NO. 15

A private developer client, with whom we regularly deal, was making a build-to-suit proposal to a prospective tenant on a piece of land that our client controlled. They asked if we would like to get involved be submitting a design-build proposal. We, of course, did, and got a local architectural firm involved with us on a speculative basis.

The design firm prepared preliminary plans on the building, and we prepared a rather detailed building cost proposal along with a preliminary construction schedule.

We then learned that not only were several other developers making proposals on other tracts of land to the same prospective tenant, (which was not bad), but also that our own developer client was taking design-build proposals from several other contractors. This turned what we thought was a pretty good prospect into an absolute shot in the dark.

Had we known all this beforehand, we would never have become involved and spent thousands of dollars preparing a cost proposal and construction schedule. I also felt bad about involving the design firm in such an exercise.

Relationships and Questions

Relationships Featured

Contract to Owner

Questions

1. Shouldn't the contractor simply have asked if there was competition?

2. Do you feel the contractor was taken advantage of by the developer, and why or why not?

3. Should the contractor be reimbursed for his time and effort? What about the architect?

CASE STUDY NO. 16

The project is an approximately $70,000,000, 40-story office building located in the hometown of Contractor A. Contractor A's president sits on the Board of Directors of the owner's firm.
The owner always excludes the president of Contractor A from all the board meetings when discussing this construction project.

The owner elected to competitively bid this project, and retain a consultant to aid him in the selection of a contractor. The owner interviews contractors to include them on the selected bidder's list. The owner went to great lengths to visit contractors who happen to be in five different locations around the country. Contractor A was the only one from the local area.

The project was to be competitively bid with a closed letting. During the bidding phase, Contractor A realized that Contractor B was just completing a $100,000,000 project as a joint venture with the consultant who had been retained by the owner. An inquiry was made. The owner responded that everything was fair and above board, and there should be no concerns about this joint venture effort.

Bids were completed and the response from the owner's consultant to Contractors A and B was that they (consultant) wanted to further discuss the project with A and B in a one-on-one interview. The interview was to be held approximately two weeks after the contractors have been given the opportunity to look at some changes that are being made to the project. Contractor A and Contractor B are led to believe that others are still possibly involved in the project. In checking, Contractor A finds that all others have been informed that they are out of the competition. The consultant, however, would not admit this information to Contactor A.

One of the complicating factors to this project was the mechanical, electrical, and exterior curtain wall subcontractor packages were all bid directly to the owner.

During this two-week study phase Contractor A and Contractor B are to negotiate with the specific "subcontractors" as directed by the consultant, and to include their numbers (sub bids) in the price so that the contractor will have one total lump sum project, including all work previously bid separately. Also, a few modifications have been made to the drawings and these are also to be included into the new bid.

Lastly, a presentation is to be made before the owner, consultant, and architect for the project as to how this project will be constructed as well as the presentation of a schedule for completion.

The two weeks go by. Presentations are made at which time Contractor A and Contractor B provide the owner with a new lump sum price including all of the above. After interviews, one week is set aside for the owner and consultant to review the proposals and make their final decision. The decision announced was that final award would be to Contractor B.

Contractor A was able to find out from the owner that on the original bid Contractor A was the low bidder. However, on the rebid their number was not successful (i.e. not low). Also, after award of the contract, Contractor A discovered that Contractor B had now established an office in the home town of the consultant, and is in fact, a tenant of the consultant.

Relationships and Questions

Relationships Featured

Contractor to Owner
Contractor to Architect/ Engineer
Contractor to Other Contractors

Questions

1. What was wrong with awarding the contract to the final low bidder—Contractor B?

2. Is there a possible conflict of interest involved here—one that could or should have disqualified Contractor B or the consultant?

3. Did the president of Contractor A neglect his duties as a director by not pointing out to the board the possible conflict of interest

between the consultant and Contractor B? How can he justify not pointing it out?

4. By withholding the fact that only two contractors remained in the running for the contract, was the consultant putting Contractor A at a disadvantage? Why is this information significant to the case?

CASE STUDY NO. 17

The project is an office building in the $8,000,000 range. It is to be negotiated, and the owner is interviewing local contractors and one from the city where the main office of the parent company (owner) is located. Interviews are to be conducted with three representatives from the local office and one representative from the parent office. Contractor B is from out of town and recently completed a $100,000,000-plus project in which the owner's representative on that project is also the person who is representing the parent company in the local interviews.

Fee proposals are requested in writing along with some other data on Friday, prior to the interviews which are scheduled for the following Wednesday. Copies of the proposals are to go to each of the four people who will be in the interview.

All contractors comply, and submit their proposals as requested on Friday. Interview are then scheduled for the following Wednesday. Interviews are conducted, and Contractor B is selected as the successful contractor.

In a post mortem with the owner, Contractor A finds out that one strange occurrence did happen. During the Wednesday interview phase, Contractor B walked into the room and cut his fee on this project by approximately 30%-- before they even started their interview. This made them low bidder, and the owner decided, even though they had outlined many factors in relation to the selection, that they would elect to make the selection based on low fee.

Relationships and Questions

Relationship Featured

Contractor to Owner

Questions

1. Do you believe all contractors were treated fairly by the owner? Why or why not?

2. Why do you think Contractor B cut its fee by 30% at the beginning of the interview? Should it have been allowed?

3. Does an owner have the right to retain the contractor they feel most comfortable with as long as costs are in line?

CASE STUDY NO. 18

The individual in question is president and owner of a building construction company which puts in place $5-$10 million worth of construction annually. He has been in the business since he started the business himself "way back" when, in the 70's or 80's. He does a certain amount of design-build work, negotiates a substantial amount of work (when the construction market was busy), but historically has done most of his work through the formal bid process. Lately, he has entered the development sector wherein he owns property and will build to suit a particular client's needs.

Notwithstanding the above, the individual also is the Chairman of the Board of Trustees of a local savings bank. He is active in this capacity, and uses his assertive personality to carry considerable weight in the board's decision making process.

[Ownership of savings banks is held by the depositors. The Board of Directors, therefore, is responsible to oversee management of the bank not only in accordance with appropriate banking laws, but also to protect the interests of the depositors/owners.]

The bank is implementing a renovation/addition in the $41.5 million range and has, after a considerable amount of internal debate, entered into a contract with their Board Chairman's construction company to do the work. There was no active consideration of other contractors.

The contract is a standard AIA format, with a Guaranteed Maximum Price with a percentage fee. It is unknown whether there is a shared savings clause or what the "reimbursables" are. It is understood that the contractor is awarding subcontracts on a competitive bid basis.

Relationships and Questions

Relationships Featured

Contract to owner

Questions

1. Is there a conflict of interest in this case?

2. If you were the contractor in question, would you have done things differently or about the same way? What if your firm was desperate for work?

3. Would your opinions on this case change if you learned the contractor was doing the work at cost—an honest gesture done to save his bank and depositors some money?

CASE STUDY NO. 19

The project in question was a Post Oak Inn, all-suite, wood frame hotel with an approximate bid price of $6 million. Closed bids were accepted from a selected list of six or seven bidders. Contractors had to name subs and prices.

Approximately four days following the bid the owner advised us that we were one of the two low bids. Before making a decision they wanted to wait for their building department's review, and to incorporate some other "changes". Six days following the bid, our drywall subcontractor called. He was upset, as the second bidder had secured his name and bid amount.

Our firm's price on the subsequent was $76,000; the second bidder's price was $35,000. Material prices alone on the revisions were $36,000! The bottom line is that we lost the job by $6,000.

Relationships and questions

Relationships Featured

Contractor to Owner
Contractor to Subcontractor
Contractor to Other Contractors

Questions

1. Is an owner acting ethically when furnishing one contractor's sub prices to another contractor? Why or why not? Can such action ever be justified?

2. If you were the original second low bidder, and were desperate for work, would you not use an owner's assistance in locating a low sub bid in order to secure a job?

3. Should the contractor ask or even expect his sheetrock sub to decline the offer of the second contractor?

4. Is it ethical for the owner to bid the changes before award?

CASE STUDY NO. 20

On a recent $14.5 million hotel expansion there was a select bid list of three pre-qualified general contractors. We were one of the three.

Our estimators spent more time on the project than any other, and had both electrical and mechanical subcontractors tour the existing facility to determine if they could meet the specifications and very tight schedule. Late-completion penalties were actual damages instead of a specific penalty-per-day.

While working with our subcontractors it was determined that they could not obtain the switch gear and compressors within the time allotted for the contractors.

When submitting our bid we stated "No contractor can deliver the switch gear and HVAC within the contract schedule and we will need to add an additional 30 days to the contract schedule. We will, however, make every effort to shorten the extension. We request a meeting with the owner prior to award of contract to discuss the problem and our bid."

The other two bidders, using the same subcontractors, did not qualify their bids and ours was rejected by the owner as non-responsive (without having the requested meeting) even though we were low bidder by a small amount.

Within two weeks of bid and award, the winning contractor was at the owner's door requesting extensions and substitutions of the questioned items. The job is not yet complete and is well over 30 days late at the present time.

Relationships and Questions

Relationships Featured

Contract to Owner
Contract to Vendor

Questions

1. Does a contractor have a duty to inform an owner of potential problems prior to contract?

2. Does the owner have an obligation to fully investigate a presented problem, before awarding the bid, because the low bidder may not have realized the problem?

3. Were the other contractors acting ethically when they were willing to enter into a contract even though they knew in advance that the switch gear and compressors could not be obtained within the allotted time?

4. Should the bidding subcontractors have advised all general contractors of the inability to meet the schedule?

CASE STUDY NO. 21

Plans and specifications for a project were distributed to at least three general contractors with a specified bid date of June 2 at noon. A few days after bids were submitted, our company was notified by the owner that we were the low bidder. Throughout our conversation with the owner we suggested ways which he could decrease the cost of his project. As the conversation drew to a close, it was understood that the owner needed to decrease the price, but was unsure what to change. However, he said he would let us know what he decided.

A week later we were informed that the owner was soliciting an additional three prices from 3 new, other general contractors—on the original plans.

We feel that the owner should have issued prints to six bidders in the beginning rather than have more bidders submit quotes when our quote is now "out in the street."

Relationships and Questions

Relationships Featured

Contractor to Owner

Questions

1. In the light of the contractors numerous suggestions for changes to the project, is the owner somewhat justified in seeking additional offers?

2. Is it premature for a contractor to offer price-cutting suggestions prior to signing a contract? Why or why not?

3. Do you believe that the owner has put the contractors at an unfair disadvantage by soliciting new offers?

4. Is the owner right in requesting bids from an additional three contractors? What if the original bids were publicly opened?

5. What if the owner incorporates the original low bidder's ideas in the rebid?

CASE STUDY NO. 22

Last month we completed a shopping center for a guaranteed maximum price, with a 50-50 shared savings cost. The risks that we anticipated in our proposal did not materialize, and we had a hefty savings to split with the owner.

But, my superintendent said, "Let's reconsider this—we lost money on the last job we did for this developer. He is not pressing us for a detailed accounting. Let's keep our mouths shut, and maybe he won't ask us for the savings." The truth is that he really did beat us to death on the last job.

Relationships and Questions

Relationships Featured

Contractor to Owner
Contractor to Contractor Employees

Questions

1. Would it be ethical to offset profits on this job against losses on a previous job?

2. Would it be O.K. to go to the owner and request that the two jobs be combined to level the gains and losses?

3. How would the contractor's decision affect his relationship with his superintendent?

4. What if the owner elected to perform an audit at a later date?

CASE STUDY NO. 23

A painting contractor gave me a lot of help when we acted as a construction manager on a design phase of an apartment project. He was kind enough to offer suggestions on types of wall coverings, paint selections, and worked up several painting budget prices. At bid time' we took sealed bids and opened them privately.

The low bidder was a few hundred dollars below the painting contractor who helped us with the estimate. We awarded to the low bidder. The painter who helped us is upset because we did not give him "the last look."

Relationships and Questions

Relationships Featured

Contractors to Subcontractors

Questions

1. Would it be ethical to allow the painting contractor who helped to have the "last look"?

2. Would it be ethical to award the contract to the painter who helped you at a higher price?

3. Did the contractor have any ethical duty to reflect the second painting bidder's pre-bid time, assistance and costs in considering bids?

CASE STUDY NO. 24

I have been out of school for eight months and I have finally "arrived." I will participate in estimating a condo project we're bidding next week.

My job will be to take quotations from painting subs. I have been instructed to indicate to each bidder who calls in a price that I have a figure which is 10% lower than the actual lowest number we have received. My boss said, "This will help us get the lowest possible number by giving them something to shoot at."

Relationships and Questions

Relationships Featured

Contractors to Subcontractors
Contractors to Contract Employees

Questions

1. Is it ethical to use "misinformation" to influence pricing?

2. How should the employee handle this situation?

CASE STUDY NO. 25

My boss has instructed me how to stay ahead on cash flow on a job by "front end loading" the pay request. Our excavation costs are estimated at $125,000—yet she has instructed me to show this on the breakdown as $250,000. The demolition is worth $75,000—she says to show $150,000. The total budget for our tower cranes is $185,000 which we pay in monthly installments—but she says "tell them we must have $185,000 immediately".

My boss says by getting as much of our overhead and fee "up front" we can keep a positive cash flow.

Relationships and Questions

Relationships Featured

Contractors to Subcontractors
Contractor to Contractor Employees

Questions

1. Is front-end loading ethical?

2. Does a contractor have a right to distribute fee and overhead as he sees fit on a lump sum job?

CASE STUDY NO. 26

This morning, I received four season tickets to all the Cincinnati Bengal's home games. The accompanying note said, "From your favorite concrete supplier, in appreciation of your past business."

I'm sure that the fact we're going to purchase 60,000 cubic yards of concrete next year had nothing to do with the fine gesture. Is it ethical to accept these tickets? I truly am a real Bengal's rooter.

Relationships and Questions

Relationships Featured

Contractors to Contractor Employees
Contractor to Vendor

Questions

1. Are gifts (lunches, tickets, etc.) just good customer relations or bribes?

2. At what level (a new car) does a gift become a bribe?

CASE STUDY NO. 27

I am a contractor who has been working with a developer for five months on a shopping center project. Two months ago it was discovered that four of his twelve acres lies in an "environmentally sensitive" area. This threatened to scuttle the entire development. My developer hired the brother of the Chairman of the planning commission as a "consultant." Lo and behold the problem went away.

It is time to start the paperwork for the building permit. I am now, however, intimately knowledgeable about the site, as well as the regulations governing the situation. I have come to believe that the project will, indeed, destroy a sensitive natural area and should not be built.

Relationships and Questions

Relationships Featured

Contractors to General Public (Society)

Questions

1. Is hiring someone (a consultant) who is close to a key decision maker ethical?

2. If this contractor is sure that the project will be built (if not by him by some other contractor) is he right to proceed with the job?

3. Should the contractor withdraw from consideration if he is troubled?

4. Does the developer have the right, ethically, to use his private property as he sees fit? What ethical right does government have to deny the owner's right to develop?

5. Is this an ethical question or a legal/political question?

CASE STUDY NO. 28

Smith Construction was one of several contractors invited to bid on the construction of an office/warehouse for Brown Equipment Company, a heavy equipment dealer.

Smith Construction's bid of $2,777,000 was low according to Brown. The architect notified all bidders after the bid opening that two bidders remained in the running: Smith Construction and Jones Construction.

Brown Equipment Comany's agent, according to court testimony, told Smith Construction if it could lower its price to $2,550,000, based on late addenda, it would have the construction contract. Smith Construction said it would do it.

Brown Equipment Company's agent then visited Jones, asked for the lower price of $2,550,000 and got it.

A third bidder, Capone Construction, told the agent it would do the work for $2,521,000. It got the contract. Smith Construction sued for breach of contract. (Capone was fifth in the first round of bidding.)

Defense attorneys contended as it was the owner's money and the owner could do whatever it wanted to, including a post bid action.

The owner of Brown Construction testified that bids were just the list price and, as owner, he could then "cut the fat out."

The judge instructed the jury to determine whether the pattern of conduct by the defendant was such that there was a contract between the parties.

The jury foreman said Brown Construction's assertion that if Smith Construction met the requested $2,550,000 price, it had the contract, was held by the jury to mean there was a contract. "We didn't like the contractor being jerked around, and felt Brown Equipment Company was very unethical," he added.

Smith Construction was awarded $154,000.

In open court afterwards, the jury said it wouldn't tolerate business being conducted in the manner in which it was in this case. Smith Construction's attorney felt the decision was a landmark, as he was unaware of any other such case on this point.

Relationships and Questions

Relationships Featured

Contractors to Owner
Contractor to Contractor Employees

Questions

1. Is a bid just list price that is subject to negotiation?

2. Is it ethical for the fifth bidder to offer a new price after he is told he is high?

3. Do you think Brown Equipment Company has ever been treated in a similar fashion by contractors buying equipment? Should that influence their actions?

CASE STUDY NO. 29

My general contracting firm was the successful bidder on a project located outside our regular operating area. Most of the sub bids used were from subcontractors with whom we had never worked. Our policy is to work only with the low bidder, and never shop or discount sub's prices. But, apparently the general practice in the area is to hold an auction among the subcontractors to award the subcontracts.

The low roofing subcontractor came to the pre-award conference and hears "We are here to carefully analyze your bid and then will inform our decision to award." To this the subcontractor responded, "Okay, I can cut 10%, but that's the best I can do."

Relationships and Questions

Relationships Featured

Contractors to Subcontractors

Questions

1. Do ethical standards change from location to location?

2. Is there anything wrong with accepting voluntary discounts?

CASE STUDY NO. 30

We are a general contractor with a policy to award subcontractors to the qualified low bidder. Recently, we bid an office building project to a private owner who came back to us after the bid with some minor cost-saving changes. The owner informed us that the two low general contractors would be pricing these changes. Because of our policy and the minor nature of the changes, I approached only the low bidder on each affected trade item. As the time to turn in the revised price neared, I began receiving calls from subcontractors I hadn't contacted who stated that they heard the job was being rebid and here was their revised price. Many of these price cuts had nothing to do with the changes, but where obvious attempts to buy the job.

We need this job. I've got to do something' or lose the job.

Relationships and Questions

Relationships Featured

Contractors to Subcontractors
Contractor to Owner

Questions

1. If it's obvious that the sub bidders are simply trying to "buy the job" should their bids be accepted?

2. Is it ethical to rebid a job with only minor changes to take advantage of market pressure?

CASE STUDY NO. 31

The developer of the new store my company is building elected not to have the Geotechnical Engineer prepare an evaluation of the subsurface conditions. It was decided to only provide the boring results to the Structural Engineer for his use in the foundation design.

Upon excavation of the basement, it was found that the water table was higher than anticipated and consequently saturated the clay layer. The Structural Engineer was relying upon this layer for adequate bearing capacity.

Our choices at this point were to have the engineer reevaluate the footing design and possibly widen them, undercut the footers, or ignore the obvious poor sub grade and keep the job on schedule. The owner told us upon contract award he didn't want to hear about any extra changes or extra costs.

Relationships and Questions

Relationships Featured

Contractor to Owner
Contactor to Architect/Engineer
Contractor to General Public (Society)

Questions

1. Since the owner doesn't want any changes, should the contractor put the footings at the design level and take the chance that the building won't settle?

2. Do statements like "I don't want to hear about changes," transfer the responsibility for poor work?

3. Who has ultimate responsibility for safety in this case and why?

4. Does this pose a legal as well as ethical problem?

CASE STUDY NO. 32

I am the scheduler for a major airport renovation which is approximately complete. Due to the loss of revenue to the various airlines during the project, the owner has stipulated a $5,000 per day liquidated damage clause.

As a result of bad weather and a slow start on my company's part, we are running two months behind. My boss, the Project Manager, continues to tell me to keep shortening the durations on all our subcontractors' activities so the owner sees the end date staying constant. The subcontractors say, No way! He says maybe we can get everyone to speed up and come out close to the completion date at the end.

I think I should publish the schedule as it truly represents the current work progress, but don't want to get fired.

Relationships and Questions

Relationships Featured

Contractors to Owner
Contractor to Contractor Employees

Questions

1. Should the subcontractors cooperate in shortening the schedule?

2. How should this employee resolve the situation with his boss?

3. Is there a better way of handling this problem?

4. What are the ethical considerations involved here?

CASE STUDY NO. 33

I was contracted by an owner of a proposed shopping center. He stated that he had been negotiating for the construction of the center with another contractor and that this other contractor had not been able to give him a price for constructing this center in keeping with his, the owner's, preconceived budget on the project. His budget has been based upon cost data furnished to him by the major tenants as the construction costs for their various stores.

The owner further informed me that since I had built many centers containing the various stores and shops anticipated in his center (and was thus knowledgeable of their costs),I had been recommended to him by some of the stores as one who could probably bring the total cost of the center in at his preconceived budget.

The owner furnished me a set of plans on this center for my "horseback" guess as to its cost. He also informed me of the amount of the other contractors' estimate.

A review of the plans, and our prior experience on similar work, indicated that we could probably construct the center for slightly less cost than the estimate of the contractor whom he had been negotiating. However, on questioning the owner we discovered that the contractor with whom he had been negotiating had worked with him over the proceeding year, without cost, in assisting him in his zoning for the project, getting approval or water main and sewer extensions for the project, and in alternates for his architect's consideration that might affect cost.

With all of the above in mind, we informed the owner that it was possible we could save him several thousand dollars under the contractor's bid price, but the work done by that contractor for him during the past year was of greater value than the savings that we could make for him. We recommended that he award the contract to the contractor with whom he had been negotiating.

Several months later we were contacted by the owner to negotiate another project without competition, and over the next ten years we built eight or ten negotiated projects with this owner

under very favorable, cooperative, and friendly relations.

Relationships and Questions

Relationships Featured

Contractors to Owner
Contractor to Other Contractors

Questions

1. Should the second contractor accept the job if he could meet the budget?

2. Should the owner pay the first contractor for his upfront work?

CASE STUDY NO. 34

At 9:00 AM on Monday, an individual enters the job trailer and identifies herself to the Superintendent as an OSHA inspector. The Superintendent says he would like to have the subcontractors' superintendents present for the "Opening Conference," and that he will go get them. Upon leaving the trailer, he finds his foreman and tells him to take his men, who are working on an unguarded floor, and move them down to work on some curbs at ground level. He then gets the subs' superintendents and the inspection proceeds. The result is that the subcontractors are fined because their men are working on an unguarded slab. The general contractor is not fined, however, because his people are not "exposed to hazards," having moved to the ground before the OSHA inspector entered the area.

Relationships and Questions

Relationships Featured

Contractor to Subcontractor
Contractor Employee to Contractor Employee

Questions

1. Is it ethical to change your operation just for an inspection?

2. Should the subs have been warned to leave the slab?

3. What is the General's responsibility in providing a safe work place for the subs?

CASE STUDY NO. 35

While putting together the bid on a job, the estimator found out that he would receive only one sub bid on plastering. When the sub bid was received, he felt the plastering contractor had taken advantage of the situation. So he "roughed up" an estimate on his own, and used a number that was about 20% lower. His reasoning was that the plastering subcontractor was only entitled to a "fair price," defined as "cost plus a fair mark-up," and if this sub bidder would not accept the job at the reduced price, there would be someone who would.

About fifteen minutes before bid time, the estimator's boss learned that their company would be the only bidder on the project. He then instructed the estimator to add 25% to the project in addition to their regular markup, because their adjusted price would still be within the Owner's budget. They would, therefore, only be taking advantage of a chance to charge "market price" instead of the "down and dirty price" they had when there was competition.

The general contractor was awarded the project for his bid price.

Relationships and Questions

Relationships Featured

Contractor to Subcontractor
Contractor to Owner

Questions

1. Is there a double standard?

2. Does the contractor have an ethical right to figure his own price for proposed sub bid work if he does not receive adequate sub bids?

3. Is the general contractor "taking advantage" of the owner's ignorance?

4. Did or does the owner have the right and/or opportunity of rejecting the low and only bidder when he receives only one bid? Does this make a difference?

CASE STUDY NO. 36

We are a reinforcing steel fabricator serving primarily general contractors. We offer a discount to our customers for payments received within a certain period after delivery of the reinforced steel. We do this because extended payment periods require that we borrow to cover our cash flow.

Our problem is a contractor who feels that the discount can be taken no matter how long it takes them to pay. They always have a reason (excuse) for the late payment; however that does not change the fact that we have to borrow money if they pay late. When we press them for payment they always blow up and accuse us of threatening our ongoing business relationship. The threat of losing their future business forces us to allow the situation to continue.

Relationships and Questions

Relationships Featured

Contractor to Vendor

Questions

1. Should the supplier continue to do business with the contractor?

2. Is it ethical to use buying power to get preferential treatment?

3. What are the ethical considerations involved here? Are these only business decisions?

CASE STUDY NO. 37

As a general contractor we have a firm policy that every vendor who performs work for us must sign our outstanding subcontract or purchase order form. While it is true that our documents are extremely one sided, we do not feel that we would unfairly enforce any of the unfair clauses.

Despite our record of fair dealing a major vendor refuses to sign our document, and accuses us of being unethical for requiring them to do so. It is our belief that this vendor must make a decision to trust us despite the documents; therefore we do not do business with them.

Relationships and Questions

Relationships Featured

Contractor to Vendor

Questions

1. Should "unfair clauses" be in contracts?

2. Should trust be a two-way street?

3. Should all prospective sub bidders be advised that the contractor uses his own subcontract form, and that they should not bid if they are unwilling to use it?

CASE STUDY NO. 38

We're a small, family run operation specializing in work related to the lumbering industry. We erect semi-permanent structures, build and especially renovate mills, and do a lot of miscellaneous work for this local industry. Our community is quite small, and I know virtually everyone who works for us. My father started the firm, and we remain the only contractor in town. Next to the mills, we're the town's largest employer.

About six months ago some environmentalists came up here to try to shut down the mills and their logging operations. They wanted to stop the logging, claiming that clear cutting is bad because it destroys ozone producing trees, causes erosion and water pollution, and ruins habitat for deer, moose, and other animals. The list went on. The loggers claimed that most forestry experts, our state's included, believe clear cutting is better than simply taking out selected older trees. Experts fighting experts, and I still don't know who is right. I do know that clear cutting is, in fact, ugly, and it does trash some very beautiful forests, at least temporarily.

The environmentalist soon began blocking the roads to the mills so workers, my employees and equipment included, couldn't get to the site. They even demonstrated outside my office and yard, blocking the entrance on one occasion, and shouting we're a puppet for the mills. A few rocks thrown at our equipment, nearly hit my mechanic. Neither the town, nor the police, nor the local jail could handle the bus loads that started coming up here, and the big city newspapers seemed to be supporting their cause. The protests got ugly, with lots of threats. Everybody in town wanted the "nuts" to go home. (Injunctions evidently did not work and I couldn't keep up with the confusion.)

Because of the demonstrators outside the mills we hadn't worked in nearly two weeks. Our subcontractors had critical completion deadlines, and were way behind schedule. The mill owner partially reopened, and demanded that we return to work. My employees desperately wanted to go back to work. They were very angry with the "outsiders". They have families to support, and knew we could not carry them much longer. The company also needed to generate some cash—ASAP! Everybody, the economy of this entire community, was hurting.

However, nobody wants blame for destroying nature or to intentionally or indirectly be responsible for anyone's injury. I honestly felt there was a very good chance somebody, one of them or one of us, was going to get hurt—seriously—if we started back up. I remember thinking that I sure didn't ask for this type of problem or want the responsibility.

After much thought I decided we were going to go back to work, and if some equipment got smashed or somebody got hurt, so be it. I called the police and told them we were coming, and I told the men to be careful and play it cool. But sure enough, somebody did get hurt. A kid from out of state got hit by one of our trucks. He had jumped in front of the truck, and the driver claims he couldn't stop. The kid lived, but we're still in court. I'm very sorry, but even though I thought this could happen I simply do not consider it my fault.

Relationships and Questions

Relationships Featured

Contractor to Owner
Contractor to Contractor Employees
Contractor to General Public (Society)

Questions

1. With nearly certain knowledge that someone was going to get seriously injured, if you were the contractor would you have sent your men back to work? Why or why not?

2. What if you thought there was only a 50/50 chance of someone getting hurt? How about one chance in 100?

3. Did the contractor have an obligation to the owner to return to the job, or was the owner being unreasonable in this instance? Why?

   a. The contractor in this case is faced with at least six factors
   b. The owner.
   c. His employees.
   d. His local community.

    e.   The environmentalists from out of state who just might be right.

    f.   His own love for the natural beauty of the area.

    g.   A need for cash to sustain a family firm.

4. Which of these, in your opinion, should be the most significant to him?

5. If you, as a contractor, found out that your work would, in fact, dramatically increase water pollution would you complete your contract? Why or why not?

6. Is compliance with a contract lesser, equal, or greater value than the safety of your workers? The safety of others?

7. Would your feelings in this case be different if you, as the contractor, had plenty of non-lumber industry related work in other parts of the state?

CASE STUDY NO. 39

We're a concrete supplier, and these are not the best of times. I've had to let some people go. Nobody has had any work, and some of our best customers are in real trouble. Two old-line contractors have shut their doors. The market here has been the worst I've seen in many years.

We have a good working relationship with nearly all of our clients-subs and generals. It's taken many years of good service and fair, competitive prices, but I'd like to think we're the best concrete company in our part of the state. We're definitely the largest. I'm active in the local contractors' association, and am on a first name basis with all of our regular customers.

Two really good size jobs finally came on the market. Feast or famine! Both call for a lot of concrete and several contracts will be let on the larger of the two. We've been asked for quotes from many of our major regulars, and from an out of state firm who has never done work in this area. The outfit plans to bid all the work on both jobs. It's no secret; everyone knows why they're in town. I checked them out through "appropriate channels", and they have a great reputation.

At the last local chapter meeting of our association three of our old line customers, who are also bidding on parts or all of the contracts, got me aside. They suggested that my prices to the out of state bidder, should he ask, should be a little higher than my quotes to local firms-those with whom we have already done business and worked with successfully.

I assured them that I planned to do just that, and I did. I view it as quoting the market price to an outsider and offering a discount to firms with a proven track record. I also view it as smart business.

A few days before the bids on the first job were due the guy from the out of state firm called me to say my quote was way out of line. He suggested that I wasn't treating him fairly simply because he wasn't a local, and asked for an honest price. I refused to discuss the matter.

Relationships and Questions
Relationships Featured

Contractor to Vendor
Contractor to Other Contractors
Questions

1. Did the vendor treat the out-of –state contractor fairly? Why or why not?

2. Were the local contractors acting ethically when they suggested that the vendor should quote a higher price to the out-of-town firm?

3. Are the best interests of the owner being served in this situation?

4. Do you think the vendor and local contractors would have acted differently if the outsider had joined the local contractors' association? Why or why not?

5. Is it an ethical or legal question when the local three contractors conspired to eliminate a competitor?

For more information about the American Institute
of Constructors please visit our website at
www.professionalconstructor.org.

American Institute of Constructors
19 Mantua Road
Mt. Royal, NJ 08061
P: 703-683-4999
info@professionalconstructor.org

www.ingramcontent.com/pod-product-compliance
Lightning Source LLC
Chambersburg PA
CBHW080541190526
45169CB00007B/2586